ON PRIVACY

ON PRIVACY

TWENTY LESSONS TO LIVE BY

LAWRENCE CAPPELLO

WORKMAN PUBLISHING | NEW YORK

Workman
Workman Publishing
Hachette Book Group, Inc.
1290 Avenue of the Americas
New York, NY 10104
workman.com

Workman is an imprint of Workman Publishing, a division of Hachette Book Group, Inc. The Workman name and logo are registered trademarks of Hachette Book Group, Inc.

Design by Galen Smith

The publisher is not responsible for websites (or their content) that are not owned by the publisher.

Workman books may be purchased in bulk for business, educational, or promotional use. For information, please contact your local bookseller or the Hachette Book Group Special Markets Department at special.markets@hbgusa.com.

Library of Congress Cataloging-in-Publication Data
Names: Cappello, Lawrence, author.
Title: On privacy : twenty lessons to protect your mind, body, and self in the twenty-first century / Lawrence Cappello.
Description: New York : Workman Publishing, 2025.
Identifiers: LCCN 2024026607 (print) | LCCN 2024026608 (ebook) | ISBN 9781523524174 (hardcover) | ISBN 9781523524198 (epub)
Subjects: LCSH: Privacy, Right of—United States. | Respect for persons—Law and legislation—United States. | Electronic monitoring in the workplace—Law and legislation—United States. | Cyberstalking--United States.
Classification: LCC KF1262 .C35 2025 (print) | LCC KF1262 (ebook) | DDC 327.44/80973—dc23/eng/20240614
LC record available at https://lccn.loc.gov/2024026607
LC ebook record available at https://lccn.loc.gov/2024026608

First Edition March 2025

Printed in the United States of America on responsibly sourced paper.

10 9 8 7 6 5 4 3 2 1

CONTENTS

WHY PRIVACY MATTERS

I wrote this for those of us who know deep down that our privacy is disappearing in the face of wondrous technological marvels, but who aren't about to throw away our phones or laptops over it because we really enjoy our phones and laptops and we use them for, well, everything.

This is a crash course on privacy for people who are equal parts paranoid and practical.

When finished, you'll be able to speak and act intelligently about your right to privacy without calling for the end of capitalism, or ignoring the importance of national security, or sounding like a conspiracy theorist posting from a bunker somewhere off the grid.

You're right to be concerned. Your privacy *is* disappearing. But it's not like we're all going to suddenly shut off our screens and go back to typewriters and spiral notebooks. So we have a problem. Because your privacy matters. It matters a lot.

There are aspects of your life that are simply nobody else's damn business. Full stop. Everybody who draws breath understands this concept on a basic primal level. The desire for privacy is a fundamental aspect of what it means to be human—it extends across all cultures and histories.

Privacy is essential to human dignity.

Privacy acts as a shield against people who would take snippets of information out of context to attack our reputations or manipulate us for their own personal benefit.

Privacy creates space for intimacy. It's a crucial building block for establishing the trust needed to forge and maintain deep bonds with our friends and loved ones. For many of us, these bonds are the very best part about being alive.

Privacy is critical to our mental health and our pursuit of happiness. It allows us to rest and recharge from the burdens of the day and remain in touch with our individuality.

Privacy protects us from becoming prisoners of our recorded past. It keeps the mistakes we made when we were young and foolish from becoming the defining moments of our entire identities, and it keeps our children from suffering that same fate.

Perhaps most importantly, the right to privacy is a big part of what it means to live in a free society. History teaches us that tyrants invade privacy to accumulate power, suppress dissent, and control their citizens. Any civilization that calls itself *free* must protect privacy rights, because privacy is a precondition of liberty.

Sadly, those of us worried about our privacy often have a hard time expressing these concerns without coming off as either conspiracy theorists, elitists, or enemies of progress. It's precisely because the average person finds the importance of privacy so hard to explain that it has become so imperiled.

So that's what this book is. Something to help you explain, quickly, why your privacy matters. And something that explains, quickly, how you can better protect your privacy while still enjoying the pleasures and conveniences of the modern world.

The Nothing-to-Hide Trap

People trying to invade the privacy of others often justify their actions through a very common and slippery argument that goes something like: "Give me access to **xyz**. If you're not doing anything wrong, you should have nothing to hide."

Privacy advocates call this the "nothing-to-hide" argument. And it's a trap. A very old one. People in power have been using some combination of these words since before the time of the Caesars. If you find yourself on the receiving end of this "logic," it means someone is trying to corner you into a vulnerable position that can be very difficult to talk your way out of.

The power behind the nothing-to-hide trap, the reason it has stood the test of time, is that at first glance it seems to make sense.

Decent law-abiding people, model employees, and faithful romantic partners don't keep dark dangerous secrets, after all. So this setup is patently aggressive. You're labeled as suspicious from the start, put on your heels, and then offered immediate salvation from further suspicion if you just surrender your privacy. Try to defend yourself, and you'll only appear more suspicious and in turn less deserving of that privacy.

But escaping the trap is easier than it seems. Look at the argument closely.

Its entire foundation is built on a false premise—that the act of having something to hide is inherently sinful or wrong. Which is really just another way of saying "all secrets are bad." Call out this deceitful and ridiculous fallacy and the trap falls to pieces.

We *all* have secrets. They're nothing to be ashamed of. In fact, many secrets impact people in positive ways. The secrets kept between romantic partners or very close friends strengthen the bonds of those relationships. The secrets shared between doctors and their patients create space for frankness and honesty when discussing deeply personal ailments. The secrets kept between business associates allow innovation to flourish and protect the intellectual property of creative people.

Secrets are a necessary part of navigating modern society. Limiting access to our personal information allows us to shape and protect our reputations and gives us a measure of control over how we're perceived by others. This is because almost all of our personal information exists in snippets. When taken out of context, the private details of our lives, these snippets, too often paint a picture of us that is skewed and not entirely true.

Unfortunately, human beings form first impressions very quickly. We base them on only a few pieces of information and with little regard for the larger complexity of people's lives.

So the desire for privacy is often less about hiding something evil and more about making sure we're not being misrepresented or misunderstood. "Personal information can reveal quite a lot about people's personalities and activities," argues Daniel Solove, a law professor who has written extensively on the nothing-to-hide trap, but "it often fails to reflect the whole person."

In the internet age, being misrepresented in a digital space is *very* common and can bring significant real-world consequences. What we say and do are highly subject to misinterpretation.

Don't misunderstand. Nobody ever gets complete control over how they're perceived. We all have a right to make our own judgments and form opinions about others, and privacy is about much more than just secrets.

But taking away our ability to limit access to our private information by saying we should have "nothing to hide" means taking away our right to play an active role in shaping our own reputations. This is a crucial right, one worth protecting. Because most of us *do* have something to hide.

And there's absolutely nothing wrong with that.

HOW TO TALK ABOUT THE NOTHING-TO-HIDE ARGUMENT:

■ **The nothing-to-hide trap is built on the false premise that all secrets are inherently sinful or wrong—which is ridiculous.** Secrets between lovers, close friends, business associates, and casual acquaintances are completely normal and often completely harmless. Limiting access to personal information gives people control over their reputations, which is an essential part of navigating modern society.

■ **Secrets are nothing to be ashamed of.** People who are not doing anything wrong still have plenty to hide. Our personal information exists in snippets, which are often taken out of context. It's not unreasonable for us to want to avoid being misrepresented or misunderstood by people making snap judgments or by those who are clearly trying to damage our reputations. In the internet age, this happens all the time. How interesting can you really be if you have no secrets?

HOW TO SAFEGUARD YOUR PRIVACY:

- All of your devices have privacy settings. Take five minutes to learn how to use them. A few quick adjustments can make a real difference in protecting your personal information. **The default settings for most of your devices are usually set to the least privacy friendly. Change them.** This isn't necessary for every new piece of tech you buy, but it should be standard practice for any device that becomes an integral part of your daily life, like your phone.

- Low-tech solutions can be more effective at protecting your secrets than high-tech fads. **Stick a $3 plastic slide on your laptop camera. It will protect your webcam from hackers better than encryption.** When locking your phone, go with an old-fashioned four- or six-digit passcode over flashy biometrics like Face ID and fingerprint readers. It's getting easier to replicate faces and thumbprints. It remains harder to hack someone's brain.

Know Where the Battlefields Are

I t's easy to see why the expression "Privacy is dead" has gotten so popular.

The modern devices that make our lives so convenient also collect alarming amounts of our personal information. And because we're now so dependent on these devices, it's not like people are going to suddenly throw away their phones, laptops, and credit cards and go back to living in the woods somewhere. Hopelessness is an entirely understandable reaction.

But privacy isn't dead. To say otherwise is to fundamentally misunderstand where the battles for privacy are being fought in the twenty-first century.

The struggle for privacy today isn't really about preventing our information from being collected. That ship has sailed. The most crucial privacy battles today concern what happens to our information *after* it's collected. And the outcomes of these battles will have a very real impact on our lives for generations to come.

Information typically travels across three stages. First it is **collected**. Then it's **processed**. Then it is **shared**. IT professionals call this journey the "data flow."

Collection. Processing. Sharing. Each of these stages is a unique privacy battleground.

Let's say you buy something online. Data about who you are, what you bought, when you bought it, where it's going, and how you paid for it are all *collected* at the time of purchase. That information is then fed into a larger historical record about what else you've bought, where else that stuff went, how else you've paid, and what all of those things combined might say about you as a customer and as a human being— this is data *processing*. That larger record is then often shared or sold to others who find such data useful, like marketing firms or credit agencies or the government—this is *sharing*. And so it flows.

That first stage: the collection stage. That's what people are usually are referring to when they say

"Privacy is dead." It's another way of arguing that it's naive to think we can ever meaningfully stem the tide of modern data collection. And they're absolutely right.

But those other two stages—the processing and sharing stages? That's where the modern privacy battlefields are located.

Privacy doesn't die the moment personal information is collected. If that were true, then cybersecurity, which is all about protecting information that has already been collected, wouldn't be a multibillion-dollar industry.

There are a number of very effective tools that can be used to protect our privacy after we've knowingly, or unknowingly, handed over our data. Turning these tools into laws and policies and regulations with teeth is what today's privacy advocates are fighting for.

One practical solution is to force companies and governments to recognize a right to **access**—which means they'd have to turn over a report of all the data they've collected about us immediately upon demand—like we already do with credit reports. Another is the right to **erasure**, more popularly known as the "right to be forgotten," which would require those who collect our data to delete all of it after a set period of time. That includes search engines such as Google. A third is the right to **restriction**, which would penalize companies harshly when they share or sell our data without express consent.

The hard truth we all need to accept is that while we should still do what we can to limit large-scale data collection, any serious effort to reverse this trend en masse is a waste of energy. What hope we have lies in shifting our attention to the next two steps in the data flow: processing and sharing.

14 Privacy is not dead. But there's definitely less of it nowadays. Stopping information collection was the goal of twentieth-century privacy advocates. They lost. Protecting our personal information *after* it's collected is the privacy battleground of the twenty-first century. The sooner we understand and accept that, the sooner we can start taking back our privacy.

HOW TO TALK ABOUT PRIVACY AND BIG DATA:

■ **The fight for privacy today isn't about stopping the collection of our personal data— it's about what happens to our data after it has been collected.** Information moves in three stages: collection, processing, and sharing. Each stage is a unique privacy battlefield. The fight over data collection is lost. The battles over processing and sharing are what modern privacy advocates are waging now.

■ People who say "privacy is dead" misunderstand where the contest over privacy is being fought in the internet age. **Privacy doesn't die the moment personal information is collected.** If that were true, then cybersecurity, which is all about protecting information already collected, wouldn't be a multibillion-dollar industry.

HOW TO SAFEGUARD YOUR PRIVACY:

- The idea of retaking control over your personal data may feel overwhelming, but it's easier than you think. For a fee, online reputation management professionals will scour the internet on your behalf to identify who has your information, get it deleted or corrected, and continue monitoring your online persona as new items pop up. It's called ORM (online reputation management). Look it up and see if it's a fit for you.

- Lots of groups say they protect the data they hold by "anonymizing" it, or removing key identifiers such as your name and date of birth. This used to work. It doesn't anymore, so don't be fooled. The tech has gotten so advanced that even anonymized data can now be traced back to specific individuals. If you have data out there you're worried about, demand it be erased. And don't be shy: Many states have laws requiring companies to delete your data upon request.

Privacy Creates Space for Intimacy

I magine you're on the best first date of your life. You're sitting across from *THE ONE*. They're amazing. They're enchanting. Everything's going perfectly. You're killing it. What does that look like to you?

The setting and the activity will vary depending on who you are, but for most people the interpersonal aspects will be very similar. The talking is easy, conversation flows effortlessly, and as the time passes you both open up and share details about your lives that you wouldn't necessarily if the date was going poorly.

Privacy creates space for intimacy. We share private details about ourselves—our secrets—to communicate to others that we're interested in forging a genuine connection. This is a fundamental ingredient of establishing trust. Our private information acts as a kind of currency in relationship building. It separates the real talk from the chitchat. And the less privacy we have, the more this currency is devalued.

Physical privacy is crucial for sexual intimacy. Even the most confident individuals need a private space to fully embrace their sexuality with someone else (for the most part). This is true as much for married couples as it is for casual flings. This kind of privacy is regarded as so important that in most places violating it is against the law.

When people feel that their privacy is respected, they're more inclined to open up and share intimate details about their thoughts, feelings, and experiences. This kind of openness creates a sense of closeness and connection that builds trust and deepens relationships.

This is no small thing. These relationships, the deep connections we form with others, may well be the very best part about being alive.

This is also why, when you share secrets given in confidence without consent, it can inflict significant emotional pain. This is the essence of betrayal—and it's the risk we take when we share our private selves with others in pursuit of deeper connections. Once privacy is compromised, it can be difficult, if not impossible, to recapture that intimacy with someone.

HOW TO TALK ABOUT THE NEED FOR PRIVACY TO PROTECT INTIMACY:

- Privacy creates space for intimacy. We share private details about ourselves—our secrets—to communicate to others that we're interested in forging a genuine connection. **Our private information acts as a kind of currency in relationship building.** The less privacy we have, the more this currency is devalued.

- Our intimate relationships, the deep connections we form with others, may well be the best part about being alive. **Without privacy, it is significantly harder to forge these kinds of bonds.**

HOW TO PROTECT INTIMACY:

- Be the kind of person who keeps other people's secrets. Don't be quick to share a juicy story about someone else for a bit of short-term attention. Be mindful of how privacy can be a building block in your relationships, and you'll attract the kinds of friends who are more likely to keep your secrets safe. Nobody likes a gossip—not in the long term, at least.

- Be a privacy-conscious friend on social media. Don't tag people in photos or share their location without their permission. Nor should you post personal information they've shared with you—like a pregnancy, or a new job, or a major new purchase—without their permission. Treat their privacy like it's your own and they'll likely do the same for you.

21

We Are Prisoners of Our Recorded Past

ong ago in Sicily, there lived a rich and powerful king. Despite his great wealth and status, he could find no joy or comfort in his life. The king was a tyrant, but not a fool. He understood that ruling with an iron fist had made him many enemies, and so he spent his days consumed by a constant fear of assassination.

One day a stranger arrived at court. Looking to advance his own position, he showered the king with compliments and remarked at how wonderful the ruler's life must be. "Since this life delights you," the annoyed king replied, "do you wish to taste it yourself and make a trial of my good fortune?" Overjoyed,

23

the speaker agreed, and was soon placed on a golden couch, given the most delicious food, and tended to by the most attractive servants.

After a while the stranger, whose name was Damocles, noticed the king had hung a razor-sharp sword from the ceiling, positioned directly above where he was lying, suspended by a single strand of hair. Damocles could find no relaxation on the golden couch—his eyes forever returning to the sword hovering above him, threatening grave harm at any moment. The king had made his point.

Something you hear often from people born in the twentieth century is that they feel lucky to have had childhoods where they could explore the world around them while making dozens of mistakes that nobody would ever learn about. That they had the freedom to do or say things that were undeniably idiotic or embarrassing without the possibility of someone later pulling out their phone to show others what happened.

Many of these people are now terrified for their children. It's an understandable fear.

Today, we are largely prisoners of our recorded past. Within a single generation our words and deeds have taken on a degree of permanence never before experienced in human history. For the first time ever, many of the mistakes we make throughout our lives will remain instantly accessible to any stranger inclined to take thirty seconds for a quick online search. The same goes for the mistakes made by our children—who are often as naive and (yes) idiotic as we were.

We can hope that this new permanence will prompt society in general to grow more accepting of the reality that everyone makes mistakes and that people should be judged on the full spectrum of their words and deeds. Sadly, the evidence suggests that the internet's memory is long and its forgiveness short. There now exists a very real danger that a snippet from our past will grow to become the centerpiece of our social reputations and follow us for the rest of our lives—hanging over us like the sword in the Damocles parable.

The trouble with our new way of life is that it ignores human beings' tremendous capacity for growth and change. This fact, that we're always changing, is why most nations have statutes of limitations built into their laws for certain crimes and debts.

Statutes of limitations are powerful things. The idea, firmly embedded in our economy and our criminal justice system, holds that after a certain period of time an unpaid debt or an unpunished crime should no longer be held against someone. It is among the purest expressions of humanity's recognition that the person standing in front of you is in many ways not the same person they used to be and should not be entirely defined by their past mistakes.

We've all done reckless things. Our children will do reckless things. And some of those reckless things, done before their brains are fully formed, will be shocking and completely out of character.

26

How do we break free? Some countries have turned to the idea of a new legal "right to be forgotten." The right to be forgotten forces search engines and websites to erase certain information tied to a person's name upon request. It's not a cure-all. Not all data about someone would necessarily be removed, and if we're being realistic, some data *should* remain in place even if someone wants it erased. Information about professional misconduct by licensed doctors and lawyers, the activities of public figures in their official roles, and critical data for public safety (like convictions of repeat offenders) are all things many argue should remain accessible for public protection and democratic transparency. We also need to consider the important role investigative journalism plays in our society and that the preservation of historical events is crucial to safeguarding our societal memory. And so search engines would have to weigh public interest before hitting the delete button.

The right to be forgotten also comes with potential dangers, and it could turn search engines like Google and its competitors into a kind of "censor in chief" against those who have unfashionable opinions.

Still, we must acknowledge the importance of exploring the right to be forgotten, particularly when it comes to children. We might not be able to stop strangers from finding all of our most sensitive moments, but burying the data deeper will add a valuable layer of protection for ourselves and our loved ones.

27

HOW TO TALK ABOUT THE RIGHT TO BE FORGOTTEN:

- Establishing a right to be forgotten is important because **for the first time ever, many of the mistakes we make in life will remain instantly accessible** to any stranger inclined to take thirty seconds for a quick online search. The same goes for the mistakes made by our children. The digital age has made us prisoners of our recorded past.

- There is now a very real danger that isolated snippets from our pasts could become the centerpiece of our social reputations and follow us for the rest of our lives—hanging over us like the sword of Damocles. **The problem with our new reality is that it ignores the fact that human beings have a tremendous capacity for growth and change.**

HOW TO SAFEGUARD
YOUR PRIVACY:

- Privacy laws are different when it comes to kids. Parents have powerful tools at their disposal to protect their children from the negative effects of online sharing. You can force companies to disclose all the data they've collected about your kids, demand it be erased, and revoke consent at any time. K–12 schools are limited in the kinds of information they can disclose about students without written consent or before notifying parents directly.

- If you suspect your child's information has been used maliciously or inappropriately, do not hesitate to contact your state's attorney general, who can guide you to the appropriate enforcement agencies. These accusations are taken very seriously, and the penalties for groups that violate children's privacy laws can be very severe.

Apathy Is Understandable

There will always be people who aren't particularly concerned about their privacy. Lots of them, in fact. And that's okay.

Some minds are wired to only see danger when it's right in front of them. Others feel so overwhelmed by the size and complexity of the issue they have come to believe that resistance is futile. Others just flat out don't give the issue much thought. Apathy can be very seductive.

But if you look closely, you'll notice these good people still have passcodes on their phones and locks on their doors. Few, if any, would be okay with having an entire year of their unfiltered internet search history sent to their families or to their coworkers.

For the most part, it's *other people's privacy* they're choosing not to care about.

So when discussing privacy rights with others, don't be rude. If you get frustrated by somebody else's apathy, remember that millions of people already care deeply about this issue and that we already have enough momentum to drive real change.

You don't change someone's mind by scolding them.

Small Data Paint Big Pictures

Cards on the table. Why should you give a damn if the world knows you bought a water bottle on Amazon? Who really cares that there's a trail of digital breadcrumbs left by the clothes you browsed last week, the shows you binged last month, and the songs you streamed most last year? What's gonna happen, exactly? You'll get more perfectly tailored recommendations for the things you love and have a more delightful online experience? Sounds terrible.

There's a saying among behavioral scientists: If you want to get away with something evil, put it inside something boring and complicated.

Every day we leave behind tiny snippets of information about ourselves. Those snippets are then collected and processed and shared in ways that few people fully understand. And yes, taken individually, these snippets are harmless.

The problem is that all your data, collected across years, is being funneled into large machines that are stitching together a far more revealing portrait of you than you ever agreed to share. Your browser history, your shopping sprees, your social media interactions—they're subject to what's called "secondary use." That's when data given for one reason is passed off to someone else for entirely different purposes. Machine algorithms using artificial intelligence then cleverly reassemble all your seemingly unrelated information and uncover hidden connections that can expose you to real danger.

What danger? In short—it lets strangers screw with your mind and shape your behavior. This fact is backed by an ocean of scientific research: Big data absolutely makes us more susceptible to manipulation. All of us. It doesn't matter how clever you are. You're not AI-powered data algorithm clever.

Here's how it works. First, they scavenge your data snippets to home in on basic personal characteristics—your age, gender, education level, income. Soon, they're able to approximate certain facts about

34

your health. And your relationship status. And sexual preferences. And political views. And how deep your social connections extend.

Then things start to get really disturbing. The algorithms don't stop at basic details. They start figuring out your psychological and communication patterns and how you'll likely react to various stimuli. This is where the dangerous part comes in. These stimuli can drastically impact your personality, mood, and cognitive abilities.

Which is all another way of saying there's a machine out there holding an encyclopedia of your personal information, doing its damnedest to figure out how best to mess with your head. And it's focused on you, specifically. You.

35

Why go to all this trouble? Because we live in an attention economy. Knowing how to keep eyeballs on screens is an immensely valuable commodity. Algorithms and big data have taught us a lot about human psychology. And as our understanding of human psychology grows and grows, algorithms become that much better at shaping human behavior.

Your "algo" does not have your best interests at heart. Your algo isn't trying to make you feel good. In fact, your algo often tries to make you angry. Because studies show that when people are angry, they tend to stay glued to their screens longer. Which helps sites generate more revenue. This isn't some random stranger trying to rile you up. It's targeted. More like a sibling or a parent who knows exactly which buttons to push. You might be the most emotionally intelligent person in the world—and yet something on your screen will get under your skin sooner or later.

But let's say you're fine with all that. You shouldn't be, but let's say you are. Even if you feel you have nothing to hide from those collecting your data, that doesn't mean you have nothing to *lose.*

More and more decisions that used to be made by people are now being made by machines. This shift comes with the risk of "algorithmic bias," a term coined by data scientists to describe unintended discrimination that happens in data analysis. And it happens all the time. Algorithms are really good at what they do. Of course they are—they're AI-powered machines. But they're created by humans who have their own biases, which can inadvertently bleed into the decisions being made by the computers.

Who gets what job. What your rent is. How much your car insurance costs. Whether or not you qualify for a certain kind of medical coverage. These results can be skewed against you even if the bias isn't intentional.

The people who collect, process, and manipulate our data aren't necessarily evil. What they are is fiercely intelligent, in possession of tremendous resources, and very good at making what they do difficult to understand.

To be alive today is to exist as two beings: our flesh-and-blood selves and our digital selves. The digital version is shaped by the sum total of data that influences our opportunities for success. In an increasingly AI-driven world, the ability to control your data directly impacts your power to shape your life as you see fit.

Privacy is a form of power that allows us to take control of our own destinies.

HOW TO TALK ABOUT PRIVACY AND SMALL DATA POINTS:

- The snippets of data we leave behind are harmless on their own, but over time they're combined to reveal deeply personal things that we never intended to share. **The algorithms that have our data aren't just trying to sell us consumer goods—they're actively trying to mess with us, often by making us angry, so that we'll engage longer with online content.** We live in an attention economy.

- More and more decisions that used to be made by people are now being made by machines. This comes with the risk of digital discrimination, which can severely limit our opportunities in life. **To be alive today is to exist as two beings: the flesh-and-blood self and the digital self. The digital version is shaped by all of our data, which then influences our opportunities for success.** In an increasingly AI-driven world, the ability to control your data directly impacts your ability to shape your life as you see fit.

HOW TO SAFEGUARD YOUR PRIVACY:

- Legally speaking, not all consent is the same. "Opt-in" consent means you have to actively give permission before someone can share your data. "Opt-out" consent means your data is automatically shared from the get-go until you revoke that permission. **Opt-in:** Hey, can we share your info? **Opt-out:** Hey, we're gonna share your info until you tell us we can't. Know that in the United States, opt-out is usually the default setting. If you want to protect your data, you're going to have to be proactive.

- Set up a separate email account for your online shopping and subscriptions (called "burner emails"). Most marketing emails have trackers embedded in them that collect data about you. A burner email will keep those trackers away from your real-life contacts and important communications and let you keep all of your junk in one place.

We Can Have Our Cake and Eat It, Too

People like to talk about privacy in terms of trade-offs. We hear things like: "You can have your privacy *or* you can have national security." And: "You can have your privacy *or* you can have your smartphone." Also: "You can have your privacy *or* you can have [insert wonderfully amazing thing you don't want to live without]." The gist is that decreased privacy is a kind of price we pay for our safety and our modern conveniences.

This zero-sum thinking has to stop. It's intellectually lazy, it's factually incorrect, and it lets others get away with exploiting our private information without being held accountable for it.

We shouldn't be asking: "Privacy or **xyz**?" The real question is: "How can we have X and still preserve privacy?" Situations are seldom binary. Life's rarely this-or-that. And when we move beyond polarized thinking we usually find a productive middle ground that can balance competing needs and address a range of challenges.

Take CCTV (closed-circuit television) surveillance cameras. When used in public spaces they can be a phenomenal tool for law enforcement officers pursuing terrorists and violent criminals. They were critical in capturing the Boston Marathon bombers, and have provided crucial evidence in numerous rape and murder convictions.

They're also super problematic for privacy. Civil liberties groups, both liberal and conservative, note that the potential for privacy abuses is straight out of George Orwell's *1984*. Still, supporters of CCTV argue that less privacy is the price we must pay for security—that we can either have these cameras and thrive *or* we can tear them down and let criminals run free.

But that's not really true. There's a third path, involving tools and techniques that can be used to protect privacy while also reaping the safety benefits of CCTV.

Some cities have data-retention laws, meaning captured video can only be stored for a limited period of time before being automatically deleted. Others require their cameras to have privacy filters that will only capture images of public spaces and not include private areas such as homes or backyards. Others use software that can blur the faces captured on camera, revealing only specific suspects in specific situations. Others have strong transparency rules that demand regular reports on the number and locations of cameras, information on who has access to the images captured, and a level of public oversight to ensure that the technology is consistent with the values and principles of the community they're meant to protect.

All of these things reduce the potential that CCTV cameras will be misused, while still giving law enforcement strong tools to keep us safe.

The same logic can be applied to online purchases. People often argue that if you want to shop on sites like Amazon, or use ride sharing, or enjoy food delivery services, then you also need to give up your data. That our data is the price we pay for these conveniences.

But again, that's not quite true. These companies often collect *wayyy* more data about you than is necessary to complete your transaction. The solution? Data minimization. It's a principle that says companies should only collect, use, and store the least amount of personal information necessary for their goals. In other words, just take what you need to get the job done. Nothing more.

When you buy something online, the seller only needs to know your name, address, and payment information. They don't need to know your entire purchase history. They don't need to know your personal preferences. They don't need to store your payment information for future transactions. And they definitely don't need to share your data with anyone who isn't absolutely required to complete your transaction.

Of course privacy comes with trade-offs. Of course it does. But those trade-offs are far less severe than those who profit from our personal data would have us believe.

Zero-sum arguments are a scam. They're mostly used by groups that apply themselves, with great focus, to extracting troves of information from ordinary citizens while framing it as some kind of natural and inevitable evolution of technology and social norms. But it isn't natural. And it isn't inevitable. We don't have to choose between surrendering our privacy and living like practical people.

The truth of it is we can have our cake and eat it, too. So be greedy. Expect both your privacy and the conveniences of the modern world. More often than not, there is a third path that can provide both.

HOW TO TALK ABOUT THE TROUBLE WITH PRIVACY TRADE-OFFS:

- Avoid zero-sum thinking when it comes to privacy. **We shouldn't be asking: "Privacy or xyz?" The question is: "How do we get xyz while still enjoying our privacy?"** The world isn't black-and-white, and this kind of thinking lets others get away with exploiting our private information without being held responsible for it.

- We don't have to choose between surrendering our privacy and living like practical people. We can have our cake and eat it, too. Demand both your privacy and the conveniences of the modern world. **More often than not, there is a third path that can provide both.**

HOW TO SAFEGUARD YOUR PRIVACY:

- You can have your email and your privacy, too. You can secure your emails with **PGP** encryption, a popular tool that scrambles your writing so that only the intended recipient can read it. PGP assigns you a unique key to lock your email, and the recipient uses their own unique key to unlock it. PGP is easy to find, easy to use, and will make your emails unreadable even if someone intercepts them along the way.

- The same goes for texts. Secure your text messages with encrypted text messaging apps. Many are available for free and require only a quick download to your phone. These apps use end-to-end encryption, have auto-delete options, and are essentially uncrackable. Secure message apps have grown immensely popular since the pandemic and allow you to text with full(ish) confidence.

47

Privacy Is Essential to Human Dignity

Toward the back of an old history book about the aftermath of the US Civil War are two maps placed side by side. They both show the same cotton plantation—one sketched before the abolition of slavery and one sketched afterward.

In the first map, a cluster of rectangles marks the quarters where the plantation's enslaved workers were forced to live. The rectangles are huddled tightly together on the page, showing the cramped and suffocating conditions faced by those who inhabited them. In the second map that cluster has vanished. The slave quarters were torn down and replaced with dozens of new rectangles, representing small homes, spread across the entire plantation, all spaced about a quarter mile apart.

The landscape tells a story: Finally free, the formerly enslaved abandoned communal living and carved out private spaces for themselves.

Privacy is essential to human dignity. It gives people the power to maintain their autonomy, to establish intimate relationships, to control their reputations, and to command their own bodies—all of which are necessary to feel respected and valued as an individual. A near-total lack of privacy is one of most degrading aspects of enslavement and imprisonment and is a tool long used by those trying to subjugate others.

Dignity is a powerful thing. Take it away from someone and you take from them a large chunk of their psychological well-being.

"Privacy is essential to the development of our identities," argues Neil Richards, a leading privacy scholar. "It can allow us to develop our political and religious beliefs" and "grants us the space to engage in self-examination and the space to play, by ourselves and with others, to work out who we are as human beings."

This connection between privacy and dignity is perhaps clearest when it comes to our bodies. Our medical ailments, our nudity, our sex lives, even going to the bathroom. We've been socially conditioned to keep these things that we do private. When that privacy is taken away, the subsequent feelings of humiliation and degradation can be severe. This is why, legally speaking, your right to prevent others from using your image and likeness without your consent technically falls under privacy law.

The drastic increase of nonconsensual pornography among young people—incorrectly called "revenge porn"—is among the newest kinds of privacy violations that impact people's dignity. These violations can be so vile, so decentering, and so traumatic that they lead some to take their own lives.

But it goes beyond just our bodies. It's also about our personal information. Imagine you're at a job interview. Most employers are forbidden by law to ask about your marital status, religion, race/ethnicity, disability status, or sexual orientation. Why? Because when this personal information is kept private, it helps protect you from discrimination while increasing the likelihood that you'll be evaluated fairly and with the respect you deserve.

Controlling your personal information gives you power over your public persona. Which is another way of saying it lets you present yourself to the world on your own terms. When you lose this control, you become more vulnerable to the judgment of others, which in turn can chip away at your agency and dignity.

Dignity is a powerful thing. It's why the desire for privacy is, and always will be, a fundamental aspect of the human condition.

HOW TO TALK ABOUT PRIVACY AS A CORNERSTONE OF HUMAN DIGNITY:

- **Privacy is essential to human dignity.** It gives people the power to maintain their autonomy, to control their personal information, and to command their own bodies—all of which are necessary to feel respected and valued as an individual.

- **The desire for privacy is a fundamental aspect of the human condition.** Just about every civilization in history has adopted some kind of privacy rules or norms. When you take away someone's privacy, you take away their dignity. When you take someone's dignity, you take away a large chunk of their psychological well-being.

53

HOW TO SAFEGUARD
YOUR PRIVACY:

- Get a VPN (Virtual Private Network). This is one of the strongest and simplest privacy moves you can make. Even if you regularly delete your browser history, your internet provider is still keeping tabs on all your web activities. And sharing that data. VPNs let you browse the internet anonymously, making it almost impossible for your service provider or anyone else to see or record what sites you visit, where you are, or who you're talking to. They can be added to your phone, laptop, and tablet. Think of them as the digital equivalent of an invisibility cloak.

- Remember that not everything needs to be online. Intelligence agencies with the best cybersecurity in the world still keep a lot of their top secret information in spiral notebooks.

54

What the Constitution Says

Bad news first: The word *privacy* doesn't appear once in the United States Constitution. The good news: Legal scholars agree that the right to privacy is very much alive in that document.

The Third Amendment prohibits soldiers from being "quartered" in your home during peacetime.

Seems like a no-brainer. The idea of being forced to host soldiers in our private homes is unfathomable today. But remember that armies are expensive. Before the Revolutionary War, the British tried to cut costs by coercing American colonists to house their troops. After the revolution was won, the memory of this great violation—of being forced to let armed government agents live in our private homes—proved so powerful the founders made sure it would never happen again.

The Fourth Amendment protects you and your property from "unreasonable searches and seizures."

Before a police officer can search you or confiscate your private things they have to jump through some legal hoops. Officers must establish probable cause, find a judge who agrees with them, and obtain a warrant specifying what they can and cannot search. And to make sure everyone involved follows these rules, any evidence obtained illegally can't be used against you and will be thrown out in court. Fourth Amendment law is complicated and there are absolutely context-dependent exceptions to these rules, but it remains a bulwark of our democracy and keeps our property and privacy safe. These same protections now extend to our phones and computers.

The Fifth Amendment protects you from self-incrimination and gives you the "right to remain silent."

Think about that. There is perhaps no more conspicuous nod to the importance of privacy in our Constitution than the Fifth Amendment. When accused of wrongdoing, we have the right to simply not speak. It's the first thing cops tell citizens when they slap on the cuffs. For centuries, those in power have compelled confessions under torture and other forms of coercion. Our right to remain silent is a powerful shield against these immoral practices. In our criminal justice system, if the government thinks you did something wrong, it's their job to prove it—not yours to assist them in making a case against you.

The writers of our Constitution took their privacy, and yours, very seriously. Most people don't know that on day one of the Constitutional Convention in 1787 the framers sealed the doors of Independence Hall to keep out the public, swore a vow of secrecy, and passed a resolution that all notes about their conversations would not be published for fifty years.

They weren't trying to deceive the American people. They believed strongly in transparency and understood that tyrants often wall themselves off from the public when making political decisions. That's why as soon as the Constitution was finished, they sent the final version to the states, where almost every detail was debated widely in newspapers and sensational ratifying conventions for months on end.

But for the Constitution to be written in the first place, titans such as George Washington, Benjamin Franklin, and Alexander Hamilton needed the freedom to voice potentially explosive ideas without fear of political retribution. The framers needed privacy to speak their minds. Without it, they probably would have failed in their task. Alan Westin, perhaps the most influential privacy scholar of the twentieth century, once said that if the debates inside Independence Hall "had been made public contemporaneously, it is unlikely that the compromises forged in private sessions could have been achieved."

There is absolutely a right to privacy embedded in our Constitution. And without privacy, that document might not have been completed in the first place.

HOW TO TALK ABOUT PRIVACY RIGHTS IN THE US CONSTITUTION:

- The word may not be in it, but our Constitution has a lot to say about privacy. The **Third Amendment** keeps soldiers from being quartered in our homes, the **Fourth Amendment** protects us from unreasonable searches and seizures, and the **Fifth Amendment** gives us the right to remain silent. It's clear the founders cared a lot about ensuring the privacy rights of American citizens.

- **The Constitution was written in private.** The framers closed the doors of Independence Hall to the public, swore a vow of secrecy, and passed a resolution that all notes about their conversations not be published for fifty years so they could voice potentially explosive ideas and speak their minds freely. Without this privacy, the Constitution probably wouldn't have been completed in the first place.

HOW TO SAFEGUARD YOUR PRIVACY:

- Be respectful when speaking to police officers, but know that you have the right to ask questions if they use evasive language. "I'm just gonna take a quick look through your bag if that's okay" or "Please go ahead and open up your trunk" are cleverly worded requests—not orders. These can be met with a polite: "Officer, I do not consent to any searches. Are you ordering me to open my [bag/trunk]?" If unsure about your rights, ask questions.

- You have the right to remain silent. If you're unsure of your legal footing, use it. There's a reason it's the first thing cops tell you when you get handcuffed. Asking questions is one thing; volunteering information is something else entirely. A simple response to probing questions about where you came from or where you're going: "Officer, I'm not going to discuss my day."

LESSON #10

Big Brother
Is Real

Oceania was a sprawling superpower, known throughout the world for its advanced technology. Within its borders lived a vast network of "telescreens"—flat tablets that looked like dull mirrors installed in every home and on every street to record the movements and interactions of its citizens. Nothing escaped their eye. The government's Ministry of Peace, of course, promised the people that the screens were there for their safety. But as time passed, anyone who criticized a government decision or whispered about the possibility of political change was either arrested or mysteriously disappeared.

The true power of the telescreens wasn't so much in what they saw, but in the knowledge that they were always watching. The lively discussions in the cafés and the markets dwindled, the passionate debates over dinner tables ceased, and the city, every evening, was filled with people who had nothing new to say. Realizing that their innermost thoughts could be turned against them, the citizens of Oceania began to guard their minds and suppress any idea that might be seen as dissent. This constant surveillance led to a culture of self-censorship, where people were afraid to voice their true thoughts or to assemble for any cause.

This is the wisdom of George Orwell's *1984*, and it's why every democracy in the world has laws on the books to protect citizens from overly invasive government snooping. The book is fictional. Big Brother is not. Big Brother is very much real.

The word *surveillance* is binary in nature—which is another way of saying it means two things at once. Think of it as a double-edged sword. Our government watches us to keep us safe, not unlike how parents keep watch over their children. But like parents, they also watch us to enforce their rules and maintain their authority. And so government surveillance is about two things. It's about protection. But it's also about power.

There are two kinds of surveillance: overt and covert. Overt surveillance is the eyes we can see.

When people know they're being watched, it creates what psychologists call a "chilling effect." Individuals will often feel as though they can't act as freely as they might under normal circumstances, which can negatively impact their sense of well-being, their creativity, and their individuality. Being watched blunts our sharp edges. It fosters monotony and conformity.

When taken too far, overt surveillance can kill free expression and undermine the integrity of democratic institutions. It makes people less willing to participate in political movements that push against the status quo by instilling a sense of fear that they might be penalized or even arrested for acting in a way that the powerful find threatening. It inhibits the right to freedom of assembly guaranteed in the First Amendment. The mass arrests that accompanied Iran's 2022 protests calling for increased women's rights are a modern example of this tactic.

Covert surveillance is the eyes we can't see. When people don't know they're being watched, the potential for governments to suppress dissent and neutralize political opposition is nearly limitless.

In the 1960s, the United States government launched surveillance operations against the Civil

Rights Movement and other organizations that were advocating for greater equality. A congressional investigation the following decade found numerous declassified documents outlining how COINTELPRO (or Counter Intelligence Program), as it was called, was highly effective in disrupting these groups. Agents illegally planted evidence on movement leaders and even attempted to blackmail Dr. Martin Luther King Jr.

Thankfully, we have a crystal ball. Those Orwellian images of high-tech government actors watching our every move are no longer the stuff of fiction. We can see, quite clearly, what the future will look like if these practices are left unchecked.

At present there are, across the world, a lot of nations that can accurately be called "advanced surveillance states." Unhampered by things like constitutions and privacy laws, many of them use AI-powered facial recognition technology, internet censorship, and data mining to monitor the activities of their citizens—from their online behavior to phone calls to physical movements.

This surveillance targets and punishes individuals who express political dissent or engage in activities that are deemed too individualistic or disloyal to the government. The chilling effect on free expression and political opposition has been devastating, contributing to the further erosion of civil liberties and human rights in ancient, beautiful societies.

This surveillance studies almost everything: shopping habits, browser history, how much time people spend playing video games. What they're eating and drinking. How much they're smoking. Who they're eating and drinking and smoking with.

Step out of line, and suddenly your debit cards won't work. Or your internet gets throttled for a month. Or your kids will suddenly be disqualified from certain schools or jobs. Or an alert will pop up in the taxi you just got into telling the driver they're not allowed to take you past a certain distance because your travel has been restricted. In some cases, they'll even take away your dog.

The fiction is now reality. The technology is here and the structures are in place. And if used improperly by an ambitious tyrant, that person would be able to manipulate and undermine the fundamental principles of our republic. We're a nation that relies on the free flow of information and the ability of individuals to make informed decisions. Eliminate that, and we're done for.

Our laws, our votes, and our collective vigilance are the only thing standing between us and an autocratic future. It is the duty of a free people to watch the watchers with unwavering persistence.

HOW TO TALK ABOUT GOVERNMENT SURVEILLANCE (WITHOUT SOUNDING LIKE A CONSPIRACY THEORIST):

- **Government surveillance is a powerful tool for social control.** Strip people of their privacy and it becomes much harder for them to organize against those in power. This is why every democracy in the world has laws on the books to protect citizens from overly invasive government snooping.

- **When taken too far, state surveillance can kill free expression and undermine the integrity of democratic institutions.** It makes people less willing to participate in political movements that push against the status quo by instilling a sense of fear that they might be penalized or even arrested for acting in a way that the powerful find threatening. It inhibits the right to freedom of assembly guaranteed in the First Amendment.

HOW TO SAFEGUARD YOUR PRIVACY:

- Your cell phone is a tracking device, plain and simple. Even if you turn off your GPS, your movements are still being recorded by local cell towers (it's called CSLI—Cell Phone Location Tracking—look it up). If you don't want to be tracked but need to keep your phone close, buy a faraday bag. These small pouches have special linings that block all signals to and from whatever is inside them. Put your phone in the bag, close the zipper, and suddenly you're invisible. Nobody needs to know where you are 24/7.

- Use a password manager. They generate complex, unique, and random passwords for all of your accounts, which makes it more difficult for hackers to crack them. And be sure to lock it with a strong password—something easy to remember, hard to guess, and that does not contain any identifiable information. Think: "MyMomWalked3Miles!"

Our Bodies
Are Sacred

Everyone has the right to decide for themselves who can see and touch their bodies. This isn't a culture-specific thing. It's a human rights thing. And this right is sacrosanct.

But then, in the early 2000s, all of our phones suddenly got super advanced and now everyone is walking around with a high-quality camera and video recorder in their pocket. This technological shift, combined with the explosive growth of online pornography into a multibillion-dollar industry, drastically altered the landscape of our bodily privacy.

Among the most troubling examples is what people often call "revenge porn": when someone shares a sexually explicit image or video with another person in confidence, only for that image to be shared without consent or posted to public platforms in a massive breach of trust.

First, it shouldn't be called "revenge porn." That term suggests the victim has done something wrong that warrants retaliation. Danielle Citron, a leading privacy attorney whose work centers on preventing this behavior and seeking justice for victims, uses the more accurate term "nonconsensual pornography," which emphasizes the true nature of the crime. Someone's right to display their body on their own terms was violated. "Revenge" implies that some wrongdoer is being punished. But the only wrongdoers in this case are the ones who exploited those images. It's violation, not vendetta.

This isn't the 1990s. Sharing explicit images is something people regularly do in relationships now. People of all ages. Even if *you* don't.

Is the choice to send a racy photo sometimes regretted by the sender in retrospect? Certainly. Are there contexts where sending that kind of pic is the exact opposite of sound judgment? Absolutely. Should people exercise common sense and great caution before sending such images? Of course they should.

Nevertheless, there is no context that will ever warrant the level of humiliation that comes from having your naked body seen by thousands of strangers without your consent. To say that someone should've known better before sending a nude photo to someone with whom they were intimate is, in the modern world of dating, the "equivalent of someone telling rape victims that their skirts were too short or that nothing bad happened to them," argues Citron.

If someone sends you a sexually explicit image of themself or consents to making a recording with you, there is absolutely nothing in their doing so that implies they also consent to your sharing it. These kinds of violations, which are growing more and more common, can have a profound impact on people's mental and physical well-being and potentially demolish their career prospects.

This is a crisis that mostly impacts women. Among the droves of websites that host nonconsensual pornographic photos, approximately 98 percent of the images are of women. Many include the victims' names or other personally identifiable information. But don't think men can't also be affected. In recent

71

years the FBI has noticed a sharp increase in incidents where victims are men and even young boys.

Refusing to share explicit images without consent is a moral and ethical duty in the modern age.

Again, so much nonconsensual pornography occurs without victims even knowing. Our courts are now grappling with an influx of cases involving victims who were secretly recorded by hidden cameras set up in their partners' homes. In South Korea, the problem grew so widespread that in 2018 tens of thousands of women marched in protest of rampant spy cam porn in places like public bathrooms. "Sextortion" is another growing problem—people being emailed images of themselves along with a threat that those images will be shared with families and employers if they don't pay up.

The specific harms to the victims of these crimes can be devastating and spread to many facets of their lives. Employers might refuse to hire someone based on explicit images discovered during background checks for fear it would hurt their brand's image. Future lovers might shy away from a long-term commitment, fearing social stigma. Some victims turn to suicide.

What can we do? There *are* digital solutions for this problem. Mass takedowns of nonconsensual content is possible. Every image and video on the internet has a unique fingerprint called a "hash." Victims can identify the hashes tied to their content and work with advocacy organizations to remove anything tied to those hashes on the internet. They can also add hashes to special do-not-post lists used by popular platforms as a preemptive strike. This won't solve the problem entirely, but with periodic monitoring it can significantly reduce the chances of unwanted content being viewed by others and help victims get their lives back.

Also, from a legal standpoint, many frontline privacy advocates believe nonconsensual pornography should be brought under the umbrella of civil rights law. This would establish clearer penalties and prosecution guidelines for offenders, provide better guidance for law enforcement, and, just as importantly, recognize the moral gravity of these violations.

HOW TO TALK ABOUT BODILY PRIVACY:

- **Everyone has the right to decide for themselves who has access to their body— who can see it and touch it.** Nonconsensual pornography can have a devastating impact on victims' mental and physical well-being and impact their job prospects. Never sharing explicit images sent by a romantic partner is a moral and ethical duty in the modern age.

- If someone sends you a sexually explicit image of themself or consents to making a recording with you, **there is absolutely nothing in their doing so that implies they also consent to your sharing it**.

HOW TO PROTECT ACCESS TO YOUR BODY:

- If you believe you might be a victim of nonconsensual pornography, there are organizations staffed with dedicated experts who can help you—some of which offer pro bono legal assistance and a host of related services. The Cyber Civil Rights Initiative (CCRI) is among the very best of these organizations.

- If you're a parent, anticipate having to discuss the dangers of nonconsensual pornography with your children. Subjects like sending images, receiving images, and what your child should do if they know someone is sharing images without consent are now part of what it means for digital-age parents to have "the talk."

It's Not All About You

The desire for privacy has a lot to do with our individuality. So it makes sense that most people, when discussing privacy, lean heavily on the language of individual rights.

But privacy is also a *societal* right—something essential to the greater health and well-being of all free civilizations. When we anchor our privacy arguments to individuals' rights, we often forget this important fact and inadvertently hurt the cause of privacy in the process.

In modern democracies, citizens vote in private. There is perhaps no clearer example of privacy's value to freedom-loving societies. Voting is our life-blood. And trying to find out how someone voted in an election without their consent is considered, in most developed countries, to be a serious crime.

This wasn't always the case. In the early days of Western democracy, the way someone voted in an election was a matter of public record. Corrupt political bosses turned this to their advantage and frequently used violence and coercion to intimidate the public. Vote against the person in power and lose, and you might also lose your job, lose your business, or face violent reprisals. The "secret ballot" we use today changed all that. It gave voters the power to express their true preferences when choosing their leaders and helped guard against widespread political corruption. Laws that protect the privacy of voting are essential to the democratic process. They're not primarily meant to protect individuals. They're meant to protect our society—to give us a more accurate understanding of what we the people really think about our leaders.

Privacy rights also allow liberal and conservative activists to push against the status quo and bring about social and political change. Political groups on both the left and the right have used these protections as essential tools for more than a century to help shape and reform their nations.

For any social, political, or religious movement to be effective its members must first hold private meetings and strategy sessions so that platforms, planks, and speeches can be hammered out, honed, and debated before they're taken to the airwaves and the streets. The civil rights organizations of the 1950s and '60s would never have been so successful if every strategy session they conducted was recorded and every sympathetic financial donor, especially allies living in very racist regions, had their identities made public (though the FBI certainly tried). Again, these rights are rooted in safeguarding freedom of expression for groups, not individuals.

The same can be said for working-class labor unions. And for religious rights organizations. And for the women's movement. And for many other engines of liberal and conservative change. Most democracies have laws that limit access to the membership rosters and donor lists held by such organizations, and the US Supreme Court has unanimously ruled that citizens have a right to privacy in their political associations. The belief that the surveillance powers of the state must be constantly kept in check is a cornerstone of what it means to live in a free country.

And so privacy is, in no uncertain terms, about much more than just our individual selves. It is a *societal* right.

The problem is that the well-meaning people who debate and defend privacy often ignore this larger social value and speak only about individual rights. This omission has consequences.

The forces that work against privacy always speak about the *greater good* of society. Safer streets. A growing economy. Counterterrorism. All of these things have clear benefits and are worthwhile pursuits.

Privacy's defenders almost always speak about the protection of *individuals*. But the other side is arguing for the needs of the many. It doesn't take a doctorate in political science to see who wins the debate. Focusing on individual rights is the weaker position. Passionate speeches won't change that fact. It's a losing strategy, plain and simple.

The goal is to find a workable balance between all of the above and the right to privacy, which is also very important. Balance. That's the goal. That's what our debates should be about. Sadly, our obsession with individual rights often messes up how these debates are framed.

People who care about the right to privacy need to understand that by defending privacy only as an individual right they are dooming their cause to failure.

Yes, privacy *is* an individual right. And individual rights are extremely valuable. But if we really want to defend the right to privacy, the strongest arguments are the ones that use the language of individual rights *and* societal rights as a kind of one-two punch. We need to think bigger.

HOW TO TALK ABOUT A SOCIETAL RIGHT TO PRIVACY:

- **Privacy isn't just an individual right—it's also a societal right that's essential to the larger health and well-being of all free civilizations.** In modern democracies, people vote in private so that they can express their true beliefs without fear of intimidation or backlash. Social and political activists have a legally recognized right to organize and strategize in private so that they can better challenge the status quo. And every democracy in the world has laws that put limits on government surveillance.

- **If we only talk about privacy as an individual right, it allows the forces that push against privacy to frame themselves as supporting the needs of the many and those defending privacy as only supporting the needs of the few.** This hurts the cause of privacy. To properly defend privacy, we need to frame it as valuable both to individuals and society.

HOW TO PROTECT PRIVACY FOR MORE THAN JUST YOURSELF:

■ Consider supporting watchdogs, organizations, or movements that advocate for privacy on the societal level. These groups are often nonpartisan and nonprofit and push collective action that can lead to significant changes in legislation and corporate behavior. These changes protect us all.

■ Never feel pressured to share your political beliefs on social media if doing so makes you uncomfortable. And from time to time, reevaluate whether you're oversharing online in general. Unless absolutely necessary (say, for your career), consider keeping your posts private to assert control over who sees them. Periodically purge old DMs. In the event your account gets hacked, there will be less for the perpetrators to find.

The Reasonable Expectation Standard

P rivacy law, like most legal fields, gets more complicated the deeper you dive. But if there's a golden principle to remember it is this: The law only protects people when they have a *reasonable expectation of privacy*.

If you claim that your former lover or your boss or the police violated your privacy, the key question before the judge or jury overseeing your case is whether or not that violation occurred in a manner that a *reasonable* person would say crossed a line.

So what qualifies as reasonable? That's where things get slippery. The law doesn't exist in a vacuum.

It is a product of societal norms and politics and technology—and those things keep changing. Which means what we as a society consider to be a "reasonable expectation of privacy" keeps changing, too.

Our homes are our sanctuaries. What happens inside our personal dwellings almost always falls under the reasonable expectation standard. Home is where we rest and recharge from the burdens of life. It's where we nurture and protect our families. And in the United States we're really big on private property as a legal concept.

But in-home digital assistants and surveillance cameras are now blurring the lines of what constitutes a reasonable expectation of privacy. In certain cases, outsiders and the police can access these devices without a warrant, like during an emergency or if the device is in plain view.

Public spaces are different. Once you leave your home, the law says reasonable people should lower their privacy expectations. When others can see and hear what you're doing, you can't quite expect them to look away or to block their ears.

Yet even in public there are exceptions. It's not reasonable to say someone violated your privacy if they see you reading this book at a coffee shop. But if they lean over to sneak a peek at the notes you're jotting down in the margins, that's crossing a line. It's not reasonable to say someone violated your privacy if they watch you walk down the street. But being followed by a stranger for hours is unreasonable and would merit police intervention.

On the ground, in the real world, these scenarios can get pretty muddy. Does your boss have the right to monitor your company emails on your company computer? Totally. Does your boss have the right to monitor your personal emails on your personal phone? Hell no. But what if it's your personal email on your personal phone but you're connected to the company Wi-Fi? See? Tricky. Judges and lawmakers then have to decide what's *reasonable* in these situations.

Usually the law seeks a middle ground. Police can absolutely search your pockets and vehicle if they have probable cause, but they can't go through your phone without a warrant. Lovers can secretly follow you without permission, but they can't place a GPS tracker on your car. Employers can monitor their workers, but typically can't monitor how those workers use their personal devices. Stores can use cameras for theft prevention, but in most states prohibit cameras in dressing rooms.

The larger lesson here is that you should never forget that what society considers *reasonable* is constantly changing. And as technologies like AI, DNA mapping, and cloud-based computing become more integrated into our daily lives, expect privacy norms to change at breakneck speed. The future is uncertain. Act accordingly.

HOW TO TALK ABOUT THE REASONABLE EXPECTATION OF PRIVACY:

- **Privacy law's golden principle: The law only protects people when they have a reasonable expectation of privacy.** If you claim that someone violated your privacy, the key question before the judge or jury overseeing your case is whether or not that violation occurred in a manner that a reasonable person in our society would say crossed a line.

- **What's considered "reasonable" keeps changing.** As innovations like AI, DNA mapping, and cloud-based computing become more integrated into our daily lives, expect privacy norms to change at breakneck speed.

HOW TO SAFEGUARD
YOUR PRIVACY:

- Never share your phone's password with anyone who isn't rock-solid. That includes friends and loved ones. It takes less than sixty seconds for someone to install an "invisible" cyberstalking app on your phone that could record and share your keystrokes, photos, videos, texts, social media activity, and locations in real time. These apps, once installed, can be entirely undetectable and impossible to remove, and nothing short of buying a new device will put you in the clear once infected.

- Working remotely from a café or pub may feel like a luxury, but be warned: Your devices are very vulnerable when connected to public Wi-Fi. Think of public Wi-Fi as a filthy gas station restroom: something best used only in an emergency. If you need to connect when out and about, use the personal hotspot option on your phone instead. Or at least use a VPN. Your connections will be much safer from prying eyes.

LESSON #14

Privacy Is Essential to Mental Health

A friend of mine—an introvert—owns a very small and very old home in Birmingham, Alabama. I love this house. The wood floors are original and creak loudly. The windows are made from leaded glass, which is known for being beautiful, blurry, and letting all of your heat out. And above her large farmhouse-style kitchen sink hangs a tasteful hand-painted sign she bought at a state fair that reads: "Home Is Where Them Fuckers Ain't."

For lots of people, one of the things that makes a house a *home* is its capacity to be a safe haven they can retreat into. Especially if there's at least one room inside they can call their own. A home is a place of privacy that allows us to escape the eyes and ears of others, cast off the pressures of the day, and be our authentic selves.

This kind of solitude is essential to mental health.

We all wear masks when out in the world. It's not so much that we're liars—more that navigating modern society involves presenting slightly different versions of ourselves as we adapt to our surroundings and the people we're engaging with. You're a slightly different person at work than you are with your friends. You're slightly different with family than you are with your romantic partner. You're slightly different with absolute strangers than you are with casual acquaintances.

Solitude gives us a break from all of that. Solitude rejuvenates our minds and is crucial for mental health because it allows us the opportunity to relax, recharge, and reflect on the events of our lives. It offers a chance for introspection and self-discovery, which in turn leads to a better understanding of ourselves and improved psychological well-being. Spending time alone, whether in a room or your car or when taking a walk, helps reduce stress and anxiety and can improve our ability to focus and be more productive. It puts us back in touch with our true selves and empowers us to establish healthy boundaries with others and foster positive relationships.

"Research has shown that privacy mediates important psychological needs, such as the ability to have a fresh start, recover from setbacks, and achieve catharsis," explains Dr. Elias Aboujaoude, clinical professor of psychiatry at the Stanford University School of Medicine. Being unable to carve out private time for yourself over the course of your week can lead to feelings of insecurity, vulnerability, and powerlessness—all of which can contribute to a range of mental health issues, such as anxiety, depression, and post-traumatic stress disorder.

That I'm writing this as the world grapples with a loneliness crisis is not lost on me. We humans are social beings. We have an inherent need to interact with others. Prolonged periods of isolation and disconnection can be disastrous for our psyches. It's well proven that the quality of our personal relationships plays a significant role in our overall happiness. This, too, is a problem that has become increasingly prevalent in the internet age, where people can effortlessly retreat to the warm, dopamine-releasing distractions provided by screens.

But the world is grappling with a privacy crisis, too. And the medical community is sounding the alarm. Physicians and researchers like Dr. Aboujaoude argue that the mental health effects of online privacy violations are now so dire that they should be viewed as a public health priority. "We do a lot of education with patients," he says, "about disease, about trauma, about strategies to restore mental well-being. Internet privacy can fall within that." These clinicians are advocating for a basic right to privacy as a means to safeguard psychological health.

In no uncertain terms: Carving out a private space for your solitude is something doctors very much want you to do.

HOW TO TALK ABOUT PRIVACY'S MENTAL HEALTH BENEFITS:

■ **Privacy gives us a break from the burdens of social interaction and is crucial for mental health.** It provides an opportunity to relax, recharge, and reflect on the events of our day. It offers a chance for introspection and self-discovery, which in turn leads to a better understanding of ourselves and improved psychological well-being.

■ Spending time alone, whether in a room, in your car, or taking a walk, helps reduce stress and anxiety and can improve our ability to focus and be more productive. **It gives us a sense of personal autonomy, empowering us to establish healthy boundaries with others and foster positive relationships.**

HOW TO PROTECT YOUR MENTAL HEALTH THROUGH PRIVACY:

- Don't be reluctant to communicate to others that you need to carve out a private space for yourself. Some may take it personally, but it's a common enough impulse that most will understand if you explain yourself clearly. Try something like: "I need some time for myself to clear my head and recharge. Please understand that it's nothing personal. I'll reach out when I'm ready to reconnect."

- Police your personal space. Watch out for "shoulder surfers"—nosy onlookers who stand too close when you're entering passwords or sensitive info. If someone is consistently invading your space, they're being rude, so speak up and be rude right back. Phone thieves often work in teams: Thief #1 shoulder surfs for your password then Thief #2 snatches your phone while you're distracted. Privacy screens, which prevent snoopers from seeing your screen at an angle, are very effective weapons against this.

It's About the Money

Although it's a wonderful book—prescient, durable, and so very disturbing—George Orwell's *1984* got some big things wrong. The government never really had to strong-arm us into putting surveillance devices in our homes. We bought those devices ourselves, eagerly, in the name of convenience and entertainment.

The most significant threats to our privacy don't come from our government. Not exclusively, at least. They come from the private companies that sell us all the devices and services that make our lives easier. These businesses devote themselves, with great energy, to collecting vast stores of our personal data in the pursuit of vast sums of money.

Shoshana Zuboff, a bestselling author and Harvard Business School professor, calls this phenomenon "surveillance capitalism." And it's the dominant business model of the twenty-first century.

Surveillance capitalism is an economic system where the product being sold isn't something tangible, such as food or clothes. The product is *you*. Your data. As much of it as businesses can get their hands on—search histories, purchase histories, likes, dislikes, and thousands of other data points. All of it collected and fed into advanced algorithms that extract insights about our behavior and preferences. It's a multitrillion-dollar industry.

The shot callers are overwhelmingly the companies that provide "free" online services, with social media outlets, search engines, and major online retailers being by far the biggest players.

Exactly why is your data such a valuable commodity? That's a question with a lot of answers. But basically, your data, when used properly, can predict the future. "The idea here is that what is being produced are predictions," says Zuboff. "Predictions of future human behavior that are then sold to markets of business customers who have an interest in what people will do now, soon, and later."

We like to imagine ourselves as unique individuals on distinct paths, but the brutal truth is that human behavior is remarkably predictable. Companies that collect vast quantities of personal data on their users hold a significant advantage over their competitors. The more data they have, the better they understand their customers and the more they can tailor their services and manipulate customer behavior.

Ever have an ad pop up on your screen for something you were just talking about, despite never doing an online search for it?

It's not that big tech is listening to your conversations through your devices' microphones. Rather, they're likely monitoring your location and the locations of your social media friends. So when a group of you all get together, like at a bar or a restaurant, they know about it. While everyone's eating and drinking and talking they're rifling through your browser histories and in the event someone just bought a particular item, or recently did a lot of searches about a particular item, they'll send the others in the group an ad for that product because it likely came up in conversation. Which, when you think about it, isn't much better than the microphone thing.

This is the modern business landscape. And it comes with a lot of upsides. Companies that can harness the power of surveillance capitalism create highly curated products and experiences that are far more likely to resonate with their customers. Without a doubt, the world is a much cooler place for consumers because of it.

But it's also a nightmare for privacy. Surveillance capitalism has been used to manipulate behavior for political gain, and it gives business leaders—whom nobody elected—unprecedented power over our personal lives. Surveillance capitalism undermines the personal autonomy of consumers and fosters an economic culture where companies are incentivized toward secondary use, sharing data given for one specific purpose with others who will use it for something entirely different. And, in the process, revealing facts about us that we never intended to share.

Western surveillance states are arguably just as powerful as the one depicted in *1984*. However, unlike that narrative, this power comes from a *partnership* between the government and corporations. These companies share our private data with government agencies all the time, and the truly disturbing part is that those agencies often don't need a warrant, since they're getting the info from a third party.

I mean, who needs CCTV cameras on every street corner when they can just ask your cell phone provider for your entire GPS location history? You don't need to be a privacy expert to understand that this is dangerous territory. If we're not careful, the numerous advantages of surveillance capitalism will be wielded to undermine the very aspects of our society we hold most dear. Things like our freedom of expression, our freedom of association, and our freedom to enjoy a full array of ideas and not live inside echo chambers where we only have access to information that aligns with our existing beliefs.

HOW TO TALK ABOUT SURVEILLANCE CAPITALISM:

- **The most significant threats to our privacy don't come from the government.** They come from "surveillance capitalism"—the practices used by the private companies that sell us all the devices and services that make our lives easier and that devote themselves, with great energy, to collecting vast stores of our personal data in the pursuit of profits.

- Surveillance capitalism is a key component of the modern business landscape. And **it comes with a lot of upsides**, including companies that create highly curated products and experiences for us to enjoy. **It's also a nightmare for privacy.** Surveillance capitalism has been used to manipulate behavior for political gain, and it gives unelected business leaders unprecedented power over our personal lives.

HOW TO SAFEGUARD YOUR PRIVACY:

- Did you not read every word of that website's ninety-page privacy policy before clicking "I Accept"? Don't worry. Nobody does. Fortunately, there are pro-privacy groups that analyze these things for free to help users understand the policies of different products and services. You can visit their websites for quick breakdowns. Among the most trusted are the Electronic Frontier Foundation (EFF) and Terms of Service; Didn't Read (ToS;DR).

- Buy your own router. A lot of internet companies give you one when they hook up your internet. But that's not really yours. It's theirs. Look at your bill—they're renting it to you. And since it belongs to them, they can collect every scrap of data you send through it. Keep your home secure. New routers may not be cheap, but you'll end up spending the same money in rental charges over the course of a year anyway.

So Make Privacy Profitable

T here's an old joke: "How many politicians does it take to screw in a light bulb? Two: one to assure us everything possible is being done while the other screws it into a faucet."

Those who doubt the ability of politicians to protect our privacy when it matters have a powerful weapon at their disposal: the profit motive. Be mindful where you spend your money by rewarding companies that prioritize privacy. Economic leverage is often the only way to create lasting change in our society. Want a more private world? Make privacy profitable.

The good news is that this trend is already well underway and global markets have taken notice. In 2020, the fastest-growing company in America

according to *Fortune* magazine was a software firm
whose entire brand is built on helping businesses con-
struct privacy plans. Many in the business world see
privacy-enhancing technology as an exciting growth
sector.

This isn't to say companies don't still make tril-
lions trading in private data. Of course they do. But
an increasing number of people now know that their
data is being sold, they have a problem with it, and the
market is responding. Things once used exclusively by
spies and the hyper-paranoid are now commonplace
in millions of American homes.

Virtual Private Networks (VPNs) provide secure,
encrypted connections between devices and the internet, making it nearly impossible for others to track
your browser history and your online activity. A recent
consumer study found two-thirds of American internet users have deployed a VPN at some point in the
last five years.

Privacy-focused web browsers. Robust password managers. Two-factor authentication services.
"Uncrackable" cell phones. Search engines that vow
never to share user info. Secure email and messaging services that boast advanced encryption. All

these things are carving out spaces in the market and becoming increasingly integral to our daily lives.

Even Apple, which at the time of writing is the planet's largest company, has jumped on the privacy bandwagon with a multiyear ad campaign under the tagline "Privacy. That's iPhone." It practically screams to consumers that buying an Apple product means your private data is safer than if you went with a competitor. Do you think they did that because they care? Or because they think it'll make them money?

If you care about privacy but have a healthy sense of pessimism, this is the most practical path to the change you want. The beauty of the economic approach is that it doesn't rely on our better angels or the outcomes of elections. It relies on self-interest and greed—which, in the end, are much more dependable. If companies believe they can get rich protecting privacy, or that they can go broke if they don't, they're more likely to strike an equitable balance between privacy and profits.

Progress is far more likely when it aligns with the self-interests of the powerful. So in a sense, we can buy our way to a more private society.

But we must tread carefully. There are pitfalls. A kind of privacy inequality could emerge where those with limited financial resources have fewer options to protect their personal information than those with means. There's also the risk that companies will mislead consumers about the privacy protections they're selling, provide a bare minimum of security just to get their brand shiny, and foster a false sense of security.

Still, we're witnessing an interesting moment. The protection of personal privacy is now, historically speaking, a more profitable enterprise than it's ever been.

We mustn't be too cynical. Perhaps our lawmakers will put aside tribal obstructionism and pass meaningful privacy legislation to better protect those they represent. But experience suggests it's just as likely that they'll either do nothing or do the bare minimum and blame their opponents when a new privacy crisis boils over.

This very recent trend of commodifying privacy may be its best hope for surviving the twenty-first century intact. It is very much in our interest to keep privacy profitable.

HOW TO TALK ABOUT MAKING PRIVACY PROFITABLE:

- Privacy is now a commodity. The market has made space for privacy-friendly products, and major companies have taken notice. **It is very much in our interest to make privacy profitable, because economic leverage is an effective way to create lasting change.**

- **Progress is far more likely when it aligns with the self-interests of the powerful.** If companies believe that they can get rich protecting privacy, or that they can go broke if they don't, they're more likely to strike an equitable balance between privacy and profits.

HOW TO PROTECT AGAINST PEOPLE MAKING MONEY OFF YOUR LACK OF PRIVACY:

- Get in the habit of doing a kind of spring cleaning for your devices. Take a day and unsubscribe from unwanted email promos. Delete any app you haven't used in a year, along with the account associated with it, and clear your browser history, cache, and cookies. This will minimize your digital footprint and help you spot unusual activity.

- Avoid autofill whenever possible. Yeah, it's super convenient, but it involves storing a lot of your personal information in places you might not want it. Take the extra second and type it out. And if you're stubborn, at least use a secure password manager with an autofill option. The same goes for browser extensions. They may save a bit of time, but are usually riddled with malicious and invasive trackers.

The Road to Hell Is Paved with Good Intentions

A lot of the things that threaten our privacy began with the best of intentions to achieve praiseworthy goals. They're the kinds of things only irrational people would argue against: speed cameras in front of elementary schools to protect kids from reckless drivers, surveillance operations targeting potential terrorist groups, digital home assistants to keep us organized. . . . These things and others like them were all born from a sincere desire to create a safer and more convenient world.

But history teaches us that these well-intentioned goals tend to shift and expand over time. Experts call it "mission creep." It happens when a privacy-invading tool gets used for purposes beyond its original aims and objectives. Whether in law enforcement, intelligence gathering, or consumer data collection, mission creep is often the reason behind the very worst privacy violations.

And it happens all the time.

One of the biggest factors that contributes to mission creep is changes in political leadership. Democracies tend to swing like pendulums, with rival parties and ideologies replacing each other every few election cycles. These swings can cause surveillance programs that were originally focused on one specific population or community to pivot to a new target audience based on the prevailing social attitudes and political preferences of those in power.

Another cause of mission creep is the failure to account for transparency and accountability when the original tech is first set up. When governments and communities and private actors install privacy-invading technology hastily—when they only see the immediate problem the tech is meant to solve and don't think a few steps ahead—they forget to put in place proper oversight, guidelines, and control mechanisms. Without these things, it's easier for a surveillance program to drift off course and be used for unrelated tasks.

The common thread running through the mission creep problem is short-term thinking. That's how a digital assistant goes from helping you with your shopping list to recording how often you say certain key words in your private conversations because the company that sold it to you is now interested. That's why six more speed cameras popped up in your neighborhood, nowhere near schools, to help generate profits for local leaders who mismanaged public funds the year prior. And it's how surveillance programs originally focused on public safety end up targeting innocent people and groups that did little more than express an unfashionable opinion.

The most famous modern example of mission creep came when Edward Snowden, a contractor for the US National Security Agency (NSA), revealed in 2013 that the agency had been illegally collecting and analyzing the phone and internet records of millions of Americans. Although originally justified as an anti-terrorism measure, it was later revealed that the NSA had been using the program to collect data on a wide range of other activities in a blatant violation of the Fourth Amendment.

So what do we do? Abolish surveillance? That would be impractical. Ultimately, we want the things we were originally promised—safety, security, convenience—before things got all screwed up. What we need are better safeguards.

The solution isn't "Don't use the tech." We should never stop embracing the benefits of new technology. The solution is "Only use the tech if we can clearly define its goals and objectives and are able to monitor it."

But we need to ensure effective oversight and accountability are in place *before* doing so. Because the only thing constant in this life is change.

114

HOW TO TALK ABOUT MISSION CREEP:

- Most of the things that threaten our privacy began with the best intentions to accomplish praiseworthy goals. **But history shows us that these goals tend to shift and expand over time, leaving us vulnerable to "mission creep," when a privacy-invading tool is used for things beyond its original aims and objectives.**

- **Mission creep must be kept in check with proper safeguards.** Rules for transparency and accountability and other guardrails should be established BEFORE using new privacy-invading tech, and that tech should be periodically monitored to make sure it hasn't expanded beyond its original purpose in problematic ways.

115

HOW TO SAFEGUARD YOUR PRIVACY:

- Don't overshare your phone number. Times are different. Oversharing used to just mean more potential spam calls, but now it carries the risk of attracting scammers and identity thieves. If you want to stay in contact with a new acquaintance, start with email or social media messaging. Being mindful of who you give your number to keeps you in control of your communication channels.

- The cloud is one of the least secure places you can keep your data. Cloud computing is a wonderful tool for businesses and for those looking to stay organized across multiple devices. Its benefits are legion. But anything you want to keep absolutely private doesn't belong there.

Governments Need (Some) Privacy, Too

There's an expression among privacy advocates: "Transparency is for government. Privacy is for people." The gist here is that privacy rights should be strong for individuals, while governments should operate openly and in full view of the people.

The idea is mostly right. Transparency in government is essential to free societies. We know from experience that when leaders close their doors to the public, corruption and injustice are quick to follow. After the Nazis were defeated, the new German government ripped the roof off the old Reichstag Building (where their Parliament meets) and installed a giant glass dome—a symbol that the state would no longer gather in the shadows but in full view of the people, in the sunlight, as a true democracy should.

It's a nice sentiment. Truly. But the reality is a bit more complicated. Privacy is absolutely necessary for certain aspects of democratic governments to function effectively. Not unchecked privacy—that would be madness. But there are situations where total transparency isn't in our best interest.

Imagine you're leading a branch of government. You're a big shot. Your intentions are pure, but your responsibilities are immense and the pressure is crushing. Thankfully, you're not doing it alone. You have experts to advise you.

You'll need your advisers to speak frankly in meetings. You'll need them tossing out bold and wild and potentially unpopular opinions without fear. Because that's the only way you'll be able to get the full range of options needed to navigate the obstacles in front of you. If your closest advisers are worried that every word they say will be recorded or written down somewhere, that every crazy idea they share is going to end up in the press or could be used against them by opportunists, they're going to self-censor. And now the range of available ideas is narrower, and your leadership much less dynamic than it could be.

Remember that the Constitution was written in private, then presented to the public *after* the hard discussions among the founders were hammered out.

For the same reasons that individual citizens need privacy to better formulate their ideas and respond to problems intelligently, our leaders also need a certain degree of privacy to do the same thing. They have to be able to engage in frank discussions without fear of petty reprisals from political enemies who care less about the truth than they do about dunking on the other team.

This principle also applies to diplomatic negotiations and peace talks. Relationships between nations are delicate things. Confidentiality is often needed for diplomats to express their views freely without fear of public backlash. The outcomes of these talks are always made public and are subject to review, but exactly how the so-called sausage was made usually remains concealed. And more often than not we're better for it.

This expectation of *some* privacy in government goes back to the earliest days of the republic. President

Washington refused to hand over records relating to a controversial treaty with England. President Adams refused to share letters about the XYZ Affair with France (they embarrassed and insulted some American diplomats—it's a great story if you haven't heard it). President Jefferson refused, multiple times, to share the details of ongoing negotiations for what eventually became the Louisiana Purchase. And if he hadn't, it might have screwed up the greatest land deal in American history. In each of these episodes, the presidents struck a balance between privacy and transparency. They all withheld information about their *process* but provided transparency in the *outcomes*.

Nobody can deny privacy is essential for national security. Loose lips sink ships. Governments hold classified information about their military, intelligence operations, and diplomatic relations. If disclosed, that information could endanger the nation and put lives at risk.

This is not an argument for some governmental-privacy free-for-all. Just a reality check. Extreme displays of government privacy should be condemned outright. The potential for abuse here is very real and is well-documented in the lessons of history. When in doubt, it's often wise to favor increased transparency in government, not less.

Remember President Nixon? He claimed (read: lied) that he couldn't release incriminating secret recordings due to "national security" and "executive privilege." The Supreme Court unanimously disagreed and forced him to share those recordings with Congress, and he resigned soon after.

But even in that famous decision (*United States v. Nixon*), the Court—whose deliberations occur in private, by the way—took pains to point out that the larger idea of privacy for presidents is important and remains valid. Nixon only lost that right to privacy because he used it to cover up acts most people consider illegal.

Transparency is for government. Privacy is for people. Yes, mostly. This is a good compass to guide our broader perspectives. But it also falls victim to the kind of zero-sum thinking practical people ought to avoid. Like most things concerning privacy, it's about balance.

HOW TO TALK ABOUT THE NEED FOR (SOME) PRIVACY IN GOVERNMENT:

- **Privacy is necessary for certain aspects of democratic governments to function effectively.** We're not talking unchecked privacy—that would be madness. But there are some situations where total transparency isn't in our best interests. Presidents Washington, Adams, and Jefferson all knew this, and it worked out better for the nation in the end. Government advisers need to be able to speak candidly with each other and voice unfashionable opinions to ensure a full range of options is made available to our leaders. Loose lips sink ships.

- That said, **extreme displays of government privacy should be condemned outright**. The potential for abuse is very real and is well-documented in the lessons of history. When in doubt, it's often wise to demand increased transparency in government, not less.

HOW TO SAFEGUARD (SOME) GOVERNMENT PRIVACY:

- At work, privacy can be a strategic asset. When colleagues and superiors know they're off the record, they tend to speak with the sort of candor that's pivotal for problem-solving and decision-making. And if you ever get the chance for anonymous feedback, take it. You'll get a much more candid assessment of your strengths and weaknesses than you would in a face-to-face review.

- Privacy education should be woven into school curricula. Teaching younger generations about the political dimensions of privacy will foster a society that is better equipped to strike an appropriate balance between privacy and transparency. That knowledge will prove useful when evaluating our leaders on Election Day.

Insist on Privacy by Design

A master architect was commanded by an emperor to build a fortress by the sea. The emperor ordered that the fortress be magnificent, but also resilient. It should be impervious to the elements and impregnable in the face of invading armies.

Most architects would have built the fortress first, to ensure its grandeur, then considered how best to defend it against the destructive power of sieges and the sea. This architect was different. A master. Before the first stone was laid she spent months studying the tides, observing the wind patterns, and consulting local warriors about siege tactics. Only after she understood all the potential dangers did she begin construction.

When the fortress was finally built, it was both an architectural marvel and utterly indestructible.

The low-tech life certainly comes with more privacy. But let's not lie to ourselves—our devices are here to stay. So if we want a future with meaningful privacy, we should take a hard look at what digital architects call "Privacy by Design" (or PbD).

Privacy by Design is an engineering principle that says privacy and security must be prioritized from the very beginning of a project's development, rather than adding them on as an afterthought once a project is done. Like in the fortress-by-the-sea allegory, tech built with PbD is built with an understanding that privacy shouldn't be an add-on—it should be baked into the DNA of all new hardware and software from the beginning.

The idea here is that since technology got us into this mess, it should be technology that gets us out. Perhaps not *out* exactly, but at least to a much healthier place.

Take text messaging. You want to text a friend about something sensitive, but you're worried the messages might come back to haunt you. PbD messaging apps use end-to-end encryption, which scrambles messages so that only your friend can read them, then auto-deletes your messages so they can't be discovered by prying eyes after the fact. The companies who sell these apps refuse to store messages on their servers, so if they ever get hacked or subpoenaed there are literally no records for anyone to find. It's not that they *won't* let others see your data. They physically *can't* let others see your data, because it no longer exists. This is what Privacy by Design looks like. Prying eyes are effectively locked out. Your risk has plummeted. The only way you're going to get compromised by what you sent is if your friend betrays you (which was always a possibility regardless of how you messaged them).

Incorporating PbD into smart home devices makes them *much* less creepy. Your digital assistant and your TV would be privacy friendly out of the box, meaning you would have to actively make an effort to share the data they collect with others. The way it works now, your home devices share your data by default. And good luck trying to find the menu that turns that off. It's buried for a reason.

PbD home devices would also have state-of-the-art security to keep outsiders from digitally entering your private dwelling and send you monthly reports outlining exactly what data was collected about you so you can make adjustments that most reflect your comfort level. Instead of your home being something out of a dystopian surveillance state, it's more like you'd have a discreet, well-trained butler who knows when to leave the room.

This. This is how we enjoy our devices while also protecting our privacy. Not a future with total privacy, but one with significantly more privacy than we'll have if we stay on our current trajectory.

Social media platforms, fitness trackers, phones, laptops—they're all ripe for PbD. And the process isn't that hard—it requires only three key steps. First, builders need to understand the potential privacy risks associated with whatever they're designing. Second, they need to prioritize privacy during development, thinking carefully about how information is collected, processed, and shared. Third, they need to monitor and audit the product over time to ensure the privacy protections are still working.

Understand the risks. Prioritize privacy. Monitor and audit. No more duct tape. No more patching holes as they reveal themselves. No more reactive thinking.

Privacy by Design is a proactive and aggressively pragmatic solution to the question of how, exactly, we get to use our laptops and credit cards and smart appliances without giving up so much of ourselves in the process.

HOW TO TALK ABOUT PRIVACY BY DESIGN:

- **Privacy by Design is an exciting engineering principle that holds privacy and security should be prioritized from the very beginning of a project's development.** Forget privacy add-ons—instead, the designers infuse privacy into the DNA of all new hardware and software.

- **Privacy by Design won't give us a future with total privacy, but it will give us one with significantly more privacy than we'll have if we stay on our current trajectory.** It's a proactive and highly pragmatic solution to the question of how we can enjoy our laptops and credit cards and smart appliances without giving up so much of our private data in the process.

HOW TO SAFEGUARD YOUR PRIVACY:

■ Device and app updates can be annoying, but they're crucial for your privacy and security. Half of system updates happen because some hacker just figured out how to penetrate your device and the update is patching that vulnerability. So update your devices regularly. Think of it like changing the oil in your car. It's a chore, but ignore it at your peril.

■ When in doubt, log out. Always lock your devices when not in use. Even if you're only stepping away for a second. Keep your digital life as secure as your physical one. Don't stay signed into accounts when you're not using them, especially on public devices. It's the digital equivalent of parking your car on the street with the keys in the ignition and the windows down.

Privacy Makes for Unlikely Bedfellows

The rapid deterioration of our privacy rights is something that speaks to liberals and conservatives alike. As a social and political issue, it carries the potential to bridge some of the seemingly limitless divide between Red and Blue.

If you take a close look at American history, you'll find the fight for privacy has made for some very strange political allies.

The right to privacy is a cornerstone of the conservative intellectual tradition. It's deeply embedded in the principles of limited government, private property, and individual autonomy held dear by the Right.

The conservative champion William F. Buckley Jr. wrote in the 1960s that "more and more it becomes plain that privacy is the key to liberty." The libertarian Senator Rand Paul has said his "vision for America" is one where there is "a government restrained by the Constitution. A return to privacy, opportunity, liberty." In 2016, the Republican National Committee announced that they "oppose any attempts by government to require surveillance devices in our daily lives" and that "no matter the medium, citizens must retain the right to communicate with one another free from unlawful government intrusion."

The right to privacy is also an intrinsic part of progressive liberalism. It is deeply connected to the principles of social equality, antidiscrimination, bodily autonomy, and freedom of assembly. The democratic socialist Senator Bernie Sanders has said publicly that he is "very, very worried about the invasion of privacy rights that we're seeing not only from the NSA and the government, but from corporate America as well. We're losing our privacy rights. It's a huge issue." The Democratic President Joe Biden wrote in *The Wall Street Journal* that "We need serious federal protections for Americans' privacy. That means clear limits on how companies can collect, use and share highly personal data" and that "these protections should be even stronger for young people, who are especially vulnerable online."

Granted, it is unwise to think politicians mean everything they say. But they *did* say these things because they thought they would resonate with voters. We often talk about the need for consensus, and privacy certainly qualifies as a bridge-building issue.

Both sides believe people should be able to exercise control over their personal information. Both sides believe people should be able to exercise control over their property and their personal spaces. Both sides believe one of the things that makes Americans American is a healthy mistrust of our government. Both sides care about the ways rapidly shifting privacy norms will impact their children and grandchildren— who may never be able to escape the consequences of their youthful mistakes.

Regardless of whether or not we think the system is rigged, it's clear that there's common ground here upon which leaders can build. What to do about the deterioration of our privacy in the face of wondrous technological marvels is a question and struggle that will largely define the next generation of Americans. No one is going to be talking *less* about privacy in the decades to come.

HOW TO TALK ABOUT PROTECTING PRIVACY WITHOUT POLITICS GETTING IN THE WAY:

- The rapid deterioration of our privacy is something that speaks to both liberals and conservatives. **As a social and political issue, it has clear common ground for both sides to build upon, and it carries the potential to make for very unlikely allies.** How, exactly, our leaders respond to this problem will play a large role in how they are remembered for generations to come.

HOW TO SAFEGUARD YOUR PRIVACY:

- When all is said and done, safeguarding your privacy boils down to your state of mind. **Cherish your privacy. Insist on it. Demand respect for it**—from those you can see, from those you can't, and, most importantly, from yourself.

ACKNOWLEDGMENTS

Were I only allowed to thank one person for this book, that person would be Katie Cacouris of the Wylie Agency. My gratitude for her efforts on my behalf is beyond words. Katie—you're the real deal.

Boiling down something as complicated as privacy into twenty short essays is a somewhat obscene act. Conversations with peers, students, and friends helped with the hard choices about where to cut and what tone to strike. Among them are Joel Allen, Hon. Benjamin Bowden, Katherine Brown, Kyle David, Josh Freeman, Amy Gajda, Rene Hadjigeorgalis, Sarah Igo, Robert D. Johnson, Micki Kaufman, Katerina Kroft, David Nasaw, James Oakes, Christian Parenti, Neil Richards, Lieutenant Andrew Smith, and Barbara Welter.

My very clever editor, John Meils, was refreshingly tough but fair. I'm grateful to him, Samantha Gil, Galen Smith, Martha Cipolla, and the entire team at Workman Publishing for their hard work and extreme competence.

And finally, my family—Browns, Cappellos, Tambascos—without whom, nothing.

REFERENCES
AND FURTHER READING

INTRODUCTION

I am hardly the first person to express ideas about "Why Privacy Matters."

Neil Richards's excellent book *Why Privacy Matters* (Oxford University Press) argues that privacy is, at its core, largely about power. For more on that point, see:

Véliz, Carissa. *Privacy is Power: Why and How You Should Take Back Control of Your Data.* Melville House, 2022.

Edward Snowden and Glen Greenwald both gave TED Talks on this theme in 2014 that have been viewed millions of times online—the former remotely from Russia via an iPad robot and the latter at their October conference in Rio de Janeiro.

LESSON #1

The Nothing-to-Hide Trap

No scholar has written more extensively on the "nothing-to-hide" argument than Daniel Solove. See:

Solove, Daniel J. "Why Privacy Matters Even if You Have 'Nothing to Hide.'" *The Chronicle of Higher Education,* 15 May 2011.

Solove, Daniel J. *The Digital Person: Technology and Privacy in the Information Age.* New York University Press, 2004.

LESSON #2

Know Where the Battlefields Are

Those interested in learning more about protecting information *after* it has been collected should consider the educational resources offered by the International Association of Privacy Professionals (IAPP), a global leader in data privacy management. See:

> Densmore, Russell, editor. *Privacy Program Management: Tools for Managing Privacy Within Your Organization.* IAPP, 2019.

LESSON #3

Privacy Creates Space for Intimacy

For more on the relationship between privacy and intimacy, see:

> Rosen, Jeffrey. *The Unwanted Gaze: The Destruction of Privacy in America.* Vintage Books, 2001.

LESSON #4

We Are Prisoners of Our Recorded Past

For one of the best books on the relationship between the right to privacy and the public's "right to know," see:

> Gajda, Amy. *Seek and Hide: The Tangled History of the Right to Privacy.* Viking, 2022.

For parents concerned about sharing details of their children's lives on social media and looking for wisdom, see:

Steinberg, Stacey. *Growing Up Shared: How Parents Can Share Smarter on Social Media—and What You Can Do to Keep Your Family Safe in a No-Privacy World*. Sourcebooks, 2020.

LESSON #5

Apathy Is Understandable

The idea of apathy being rooted in problems concerning "other people's privacy" is mentioned often in this seminal book by Alan F. Westin, considered by many experts to be the most important privacy scholar of the twentieth century. See:

Westin, Alan F. *Privacy and Freedom*. Atheneum, 1967.

LESSON #6

Small Data Paint Big Pictures

For more on big data's connection to discrimination and algorithmic bias, consider Safiya Umoja Noble's work on this and other subjects, which earned her the prestigious MacArthur Fellowship (often called the "Genius Grant"). See:

Noble, Safiya Umoja. *Algorithms of Oppression: How Search Engines Reinforce Racism*. New York University Press, 2018.

LESSON #7

We Can Have Our Cake and Eat It, Too

For an excellent argument on the need to move away from binary thinking about privacy, see:

Solove, Daniel J. *Nothing to Hide: The False Tradeoff Between Privacy and Security*. Yale University Press, 2011.

LESSON #8

Privacy Is Essential to Human Dignity

If you're looking for an argument that brilliantly captures the relationship between privacy and human dignity, see:

Richards, Neil. *Why Privacy Matters*. Oxford University Press, 2021.

LESSON #9

What the Constitution Says

The constitutional implications of the right to privacy can be found in many places, including within this thoroughly researched book. See:

Smith, Robert Ellis. *Ben Franklin's Web Site: Privacy and Curiosity from Plymouth Rock to the Internet*. Privacy Journal, 2004.

Westin quoted in this chapter from *Privacy and Freedom*.

LESSON #10

Big Brother Is Real

For arguably the greatest dystopian novel of all time, see:

Orwell, George. *1984: 75th Anniversary*. Berkley, 2003.

For more on COINTELPRO, from Pulitzer Prize–winning author David J. Garrow, see:

Garrow, David J. *The FBI and Martin Luther King, Jr.: From "Solo" to Memphis*. W. W. Norton & Company, 1981.

LESSON #11

Our Bodies Are Sacred

Danielle Citron is a giant in the field of "nonconsensual pornography" scholarship and advocacy. See:

Citron, Danielle Keats. *The Fight for Privacy: Protecting Dignity, Identity, and Love in the Digital Age*. W. W. Norton & Company, 2022.

For a groundbreaking work on the relationship between privacy, feminism, and autonomy that remains an important touchstone for legal scholars, see:

Allen, Anita L. *Uneasy Access: Privacy for Women in a Free Society*. Rowman & Littlefield, 1988.

To read an excellent historical exploration of the relationship between privacy and sexuality, see:

Garrow, David J. *Liberty and Sexuality: The Right to Privacy and the Making of Roe v. Wade*. Macmillan Publishers, 1994.

LESSON #12

It's Not All About You

For more on the relationship between "individual" and "societal" arguments for privacy in the history of American law, see:

Regan, Priscilla M. *Legislating Privacy: Technology, Social Values, and Public Policy*. The University of North Carolina Press, 1995.

LESSON #13

The Reasonable Expectation Standard

What counts as "reasonable" in any society has much to do with its cultural history. For among the very best cultural histories of American privacy, see:

Igo, Sarah E. *The Known Citizen: A History of Privacy in Modern America*. Harvard University Press, 2018.

LESSON #14

Privacy Is Essential to Mental Health

To thoroughly learn about the mental health implications of privacy and the "digital self," see:

Aboujaoude, Elias. *Virtually You: The Dangerous Powers of the E-Personality*. W. W. Norton & Company, 2011.

LESSON #15

It's About the Money

For the eye-opening bestseller, see:

Zuboff, Shoshana. *The Age of Surveillance Capitalism: The Fight for a Human Future at the New Frontier of Power.* PublicAffairs, 2019.

Zuboff quoted in this chapter in an interview with *New York Magazine*:

Kulwin, Noah. "Shoshana Zuboff on Surveillance Capitalism's Threat to Democracy." *New York Magazine*, 24 Feb. 2019.

LESSON #16

So Make Privacy Profitable

Those interested in learning more about privacy-enhancing technologies (PETs) should explore the work of privacy advocacy organizations such as the Electronic Frontier Foundation (eff.org), the Future of Privacy Forum (fpf.org), and the Electronic Privacy Information Center (epic.org).

LESSON #17

The Road to Hell Is Paved with Good Intentions

For more on surveillance and "mission creep," see:

Lyon, David. *Surveillance Society: Monitoring Everyday Life.* Open University Press, 2002.

Parenti, Christian. *The Soft Cage: Surveillance in America from Slavery to the War on Terror.* Basic Books, 2003.

LESSON #18

Governments Need (Some) Privacy, Too

To read in very accessible language about the importance of privacy for executive political function in the Early Republic period in the US, see:

Berkin, Carol. *A Brilliant Solution: Inventing the American Constitution*. Harcourt, 2002.

For more on the origins and importance of executive privilege among the anti-communist hysteria of the twentieth century, see:

Schrecker, Ellen. *Many Are the Crimes: McCarthyism in America*. Little, Brown and Company, 1998.

LESSON #19

Insist on Privacy by Design

The International Association of Privacy Professionals (IAPP) has published extensively on Privacy by Design (PbD). Their website (iapp.org) is an excellent resource for the latest on PbD.

LESSON #20

Privacy Makes for Unlikely Bedfellows

Buckley quoted in this chapter in:

Buckley, William F., Jr. "What We Need Is a Law." *National Review*, 1 June 1965.

Senator Rand Paul quoted in his campaign launch for the 2016 presidential election in Louisville, KY, 7 April 2015.

The Republican National Committee quoted in their "2016 Republican Platform."

Senator Bernie Sanders quoted from NBC's *Meet the Press*, 31 May 2015.

President Biden quoted in:

Biden, Joe. "Republicans and Democrats, Unite Against Big Tech Abuses." *The Wall Street Journal*, 11 Jan. 2023.

OTHERS ON PRIVACY

"Nothing was your own except the
few cubic centimeters inside your skull."

—GEORGE ORWELL,
1984

"The right to be let alone is indeed
the beginning of all freedom."

—WILLIAM O. DOUGLAS
(US Supreme Court Justice)

"Arguing that you don't care about the right to
privacy because you have nothing to hide is no
different than saying you don't care about free
speech because you have nothing to say."

—EDWARD SNOWDEN

"A life spent entirely in public, in the presence
of others, becomes, as we would say, shallow."

—HANNAH ARENDT,
The Human Condition

"Three may keep a secret,
if two of them are dead."

–BENJAMIN FRANKLIN

"A secret's worth depends on the people from
whom it must be kept."

–CARLOS RUIZ ZAFÓN,
The Shadow of the Wind

"It's my impression that these machines may
know too damn much."

–US CONGRESSMAN JAMES OLIVER
when asked, in the early 1960s,
about the growing popularity of computers
in government recordkeeping